An
Apiary
Guide
to
Swarm
Control

by Wally Shaw

Second Edition (Revised)

Northern Bee Books

I0116966

An Apiary Guide to Swarm Control
© Wally Shaw

All rights reserved. No part of this publication may be reproduced, stored in a retrieval system, transmitted in any form or by any means electronic, mechanical, including photocopying, recording or otherwise without prior consent of the copyright holders.

ISBN 978-1-914934-36-0

Second Edition (Revised)

Published by Northern Bee Books, 2022
Scout Bottom Farm
Mytholmroyd
Hebden Bridge HX7 5JS (UK)

With the agreement of Cymdeithas Gwenynwr Cymru
The Welsh Beekeepers Association

Design and artwork by DM Design and Print

An
Apiary
Guide
to
Swarm
Control

by Wally Shaw

Second Edition (Revised)

Contents

An Apiary Guide to Swarm Control

Introduction

The beekeeper needs to understand that swarming is simply reproduction. All the complex, integrated behaviour that occurs within a honey bee colony has evolved simply to improve its chance of successful reproduction. Most beekeeper's primary interest is just one aspect of this behaviour, namely the collection of nectar and the storage of honey. In order to produce the maximum amount of harvestable honey, the beekeeper seeks to create large colonies but also to prevent them from achieving their natural destiny through swarming – so there is an inherent conflict involved. The only way of addressing this is through swarm control.

In their book entitled `**Bait Hives for Honey Bees**`, Seeley and Morse state that, `**Mature colonies have a natural urge to swarm each year unless weakened by disease or mismanagement**`. So perhaps we should not be surprised or regard it as dysfunctional when colonies swarm. Most beekeeping books understate swarming but its control is vital if good honey crops are to be consistently obtained. There is no doubt that swarm control is simultaneously the most important and most difficult aspect of colony management.

The Booklet (or Field Guide)

As the name suggests, this booklet is primarily about practical swarm control; what to do when you are out there at the hive-side faced with making a decision about management to try and control the swarming impulse. It contains a series of diagrams that can be used as a guide to management and can even be taken out into the apiary for reference if required. Because no two hives are ever the same, the diagrams need to be applied flexibly as they show generalised examples but it is the underlying principle that matters – how the management interacts with the natural behaviour of the colony. The accompanying text has been kept to a minimum with (hopefully) just enough to explain the manipulations involved and their function.

For the sake of simplicity, all the diagrams assume the use of a Modified National hive with 12 self-spacing (Hoffman) frames. However, the same management techniques can be applied to all types of movable frame hive whatever their size, shape, number and spacing of frames. The frames shown in the diagrams are colour coded according to their contents; frames containing brood are coloured **red**, honey and pollen **yellow**, drawn frames with no contents **black,** dummy board**s blue** and un-drawn frames (containing a sheet of foundation) are shown as a **thin black** line. In the real world the frames within a hive usually have a mixture of contents so the colour coding refers to the **primary contents** (the dominant characteristic) of the frame.

Most diagrams show a standard hive configuration of a brood and a half, ie. one deep and one shallow box beneath the queen excluder (the brood area), which seems to be about the right amount of space for near-native (locally adapted) bees in Wales. Other hive configurations (eg. single deep, double deep or 12x 14) require appropriate modification of the methods described in this booklet and some options may be limited or even impossible.

Types of Swarm Control

From the point of view of practical management, swarm control can be divided into two distinct parts with a clear (biological) threshold between them which is **when the colony starts queen cells**.

1) **Pre-emptive swarm control** – the type of management that can be used before queen cells are present in the hive (to try to prevent their initiation).

2) **Re-active swarm control** – the type of management that can be used when queen cells are produced (to prevent the issue of swarms).

Some beekeepers seem to have misunderstood the clear distinction between these two types of management practices and applied pre-emptive management after queen cells have been started in the hope that it would make them go away. It is a vain hope I am afraid and there is very little chance this will cause the colony to change direction.

PART 1 –
PRE-EMPTIVE SWARM CONTROL

The management activities involved in pre-emptive swarm control are **multi-purpose** and not just about preventing queen cells being started. They double-up as good beekeeping practice which aims at:-

- Promoting a large colony capable of collecting a large crop of honey.

- Systematic renewal of brood combs (particularly important for disease prevention).

- Queen replacement and making increase (when required).

Unfortunately, producing a large colony and maintaining it over the period during which a nectar flow may occur (3 months or more) can create the very problem the beekeeper is trying to avoid. This is because the larger the colony and the longer it is in that condition the more likely it is to swarm. Good pre-emptive management always delays swarming but may not prevent it happening eventually. Late season swarming is particularly frustrating because it can compromise the potential for honey production during the main flow in July (or later). Two ways of dealing with this particular problem can be found at the end of the booklet **(Section 2.5)**.

Triggers for Swarming

The triggers for swarming – the means by which the colony recognises when it is a good time to swarm and is most likely to be successful – are multi-factorial and a mixture of internal and external conditions.

Internal (within the hive)

- Size of the colony, space for the queen to lay, brood nest congestion, brood nest maturity and (possibly) the age of the queen.

- Space for nectar processing and honey storage.

- Production and/or distribution of queen substance (thought to be the main mechanism).

External

- Time in the season - the swarming urge is at a peak in May and June (when it is most likely to be successful) and declines thereafter.

- Weather – an underrated factor as interludes of poor weather (with little flying time) often precipitate swarming.

The beekeeper can to some extent control internal conditions through management of the hive but can do nothing about the external factors. It follows that pre-emptive swarm control is mostly about management of the brood area.

The main management techniques by which the beekeeper can control hive internal conditions are:-

1. Comb management
2. Box management
3. Brood relocation
4. Spitting colonies

1.1 Comb Management

The aim here is to ensure that as many as possible of the frames below the queen excluder (the brood frames) are actually used for the production of brood. During the main season, stores of honey and pollen in the brood area should be kept to a minimum. Quite early on in the season, when the colony has not yet attained its full potential size, it is the bees' instinct to create a ceiling of capped honey. These (close-to-hand) stores are a form of insurance against adverse conditions which the bees are reluctant to uncap and make cells available for the queen to lay. When this ceiling is in place the only way the brood nest can be extended is in a downward direction, so the aim is to have brood in contact with the queen excluder over as much of its area as possible. The bees will still create a honey ceiling but it will be in the first super – which is where the beekeeper wants it.

Comb management may involve moving existing frames within the brood area in order to provide space for the queen to lay but the main activity is removing old or defective frames and getting new frames of foundation drawn. It is good beekeeping practice to replace brood frames on roughly a 3 year cycle, so that means an average of 3-4 frames/year. The use of foundation is thought to be an additional disincentive to swarming by simulating brood nest immaturity and diverting bees to the task of wax making.

However, in order to be successful (get frames drawn quickly) and not damage existing brood by chilling, the introduction of foundation must be done at the right time and in the right position in the hive. Early in the season, in order to maintain brood nest integrity, foundation **must** be introduced on the edge of the brood nest - so that it becomes the next frame to be drawn if the brood nest is to expand. Later, when the colony is crammed with bees, foundation can be interleaved with brood frames. **Never** put foundation next to the hive wall because in this position it will only be drawn as a last resort – and then usually badly.

Figure 1 illustrates where foundation should be placed early in the season (left) and later in the season (right) in both deep and shallow brood boxes. Using a two-box system (ie. brood and half or double brood) the placement of foundation in the upper box is much less critical because of the warmth coming up from below. Single box users (standard deep or extra deep 12x14) should follow the procedure as shown for the lower box.

Figure 1 – positioning of foundation in a hive on brood and a half

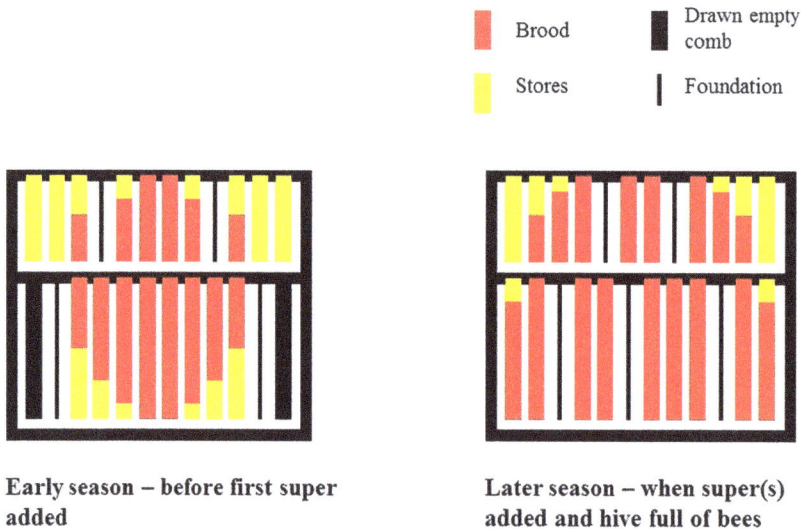

Brood
Stores
Drawn empty comb
Foundation

Early season – before first super added

Later season – when super(s) added and hive full of bees

When introducing foundation to a hive the beekeeper should understand that the colony **MUST** have an **immediate need** for more comb, either to extend the brood nest or for storage, because a colony does not do 'speculative' comb building'. There **MUST** also be a nectar flow (or the beekeeper must provide one by feeding) because bees do not use stored honey for wax-making. For further information see also WBKA booklet on *'Comb Management'*.

An alternative (less radical) approach to increasing the number of frames available for the queen to lay in the half brood is to move individual frames up into the first super When there is a good nectar flow early in the season the colony will often use any available cells in the half brood to store nectar which, particularly when it has been capped, becomes a permanent fixture and will prevent their use for brood for the rest of the season. This can be rectified by moving such frames up into the first super and replacing them with frames of foundation inter-leaved with frames that contain brood (**see Figure 2**). If the honey super has only recently been added and is not yet in full use, ie. it contains few bees, nectar bearing frames from the half-brood can be placed in the middle of the box to 'seed' it. This will encourage the bees to move up and make more use of it. Frames that contain the remains of winter stores, which may be substantially sugar syrup, should be removed altogether and used to feed other colonies in need. This is to prevent contamination of honey crop with syrup. Having been walked on by the bees for 6-7 months, such frames can be easily identified by having dark-coloured (dirty) cappings.

Figure 2 - an alternative method of maximising the use of the shallow brood box

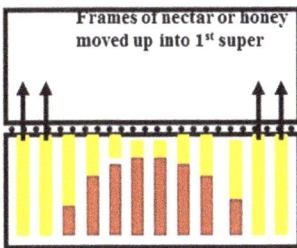

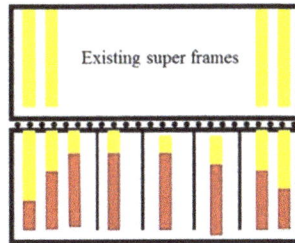

Brood

Stores

Foundation

Frames of nectar or honey moved up into 1ˢᵗ super

Existing super frames

1.2 Box Management

This is only applicable to beekeepers who use a two-box system; either brood and a half or double brood. As with comb management, the aim is to have as much as possible of the brood nest in contact with the queen excluder thus avoiding a honey ceiling.

Hive on Brood and a Half

The need for box management depends crucially on the position in the hive that the brood nest develops at the beginning of the season. If the nest is high then nothing needs to be done because the queen is free to lay down and use as much of the available space as she needs **(Figure 3a)**. A nest developing in the middle of the hive will, as the season progresses, result in a honey ceiling in the upper box and this may restrict the ultimate size of the brood nest. In this situation the boxes can be swapped – placing the shallow brood box beneath the deep **(Figure 3b)**. When the brood nest starts low then the upper box will quickly become a honey store (a super) and the brood nest size will be restricted. In this case the best option is to place the existing shallow brood above the queen excluder (make sure the queen is not in it) and introduce a new shallow brood box containing drawn frames beneath the deep brood **(Figure 3c)**. If drawn comb is not available then placing a box of foundation under the existing brood nest is the last place it will get drawn. In fact, it is unlikely to be drawn before the colony has taken the decision to start queen cells. The only option is to put the box of foundation on top of the deep brood box and below the queen excluder where it will readily be drawn. Then, before it is has accumulated a significant amount of sealed honey, to move it to bottom position. In a good nectar flow this could happen quite quickly and the beekeeper needs to catch the moment.

Figure 3a – initial brood nest in high position mostly in shallow brood box – no management required

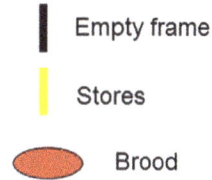

	Empty frame
	Stores
	Brood

RESULT - the brood nest naturally extends down to occupy as much of the deep box as required

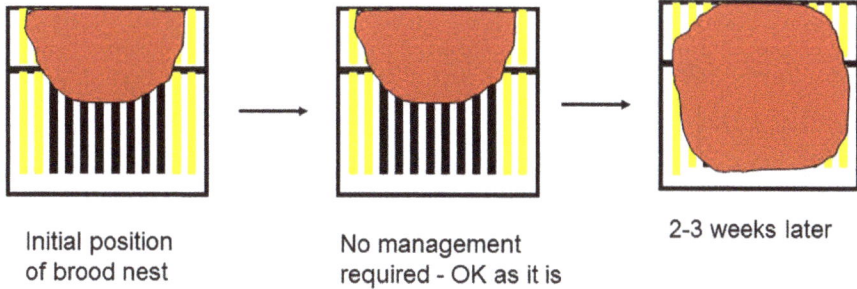

Initial position of brood nest

No management required - OK as it is

2-3 weeks later

Figure 3b – brood nest in mid-position between deep and shallow brood boxes – re-position shallow box under the deep

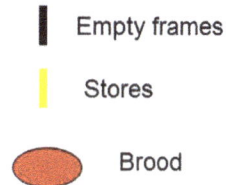

	Empty frames
	Stores
	Brood

RESULT – any honey in the shallow box is removed, comb becomes available to the queen and the integrity of the brood nest is quickly re-established

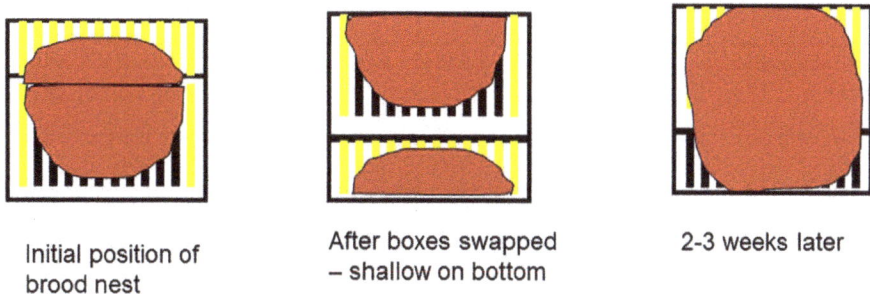

Initial position of brood nest

After boxes swapped – shallow on bottom

2-3 weeks later

Figure 3c – brood nest in low position mostly in deep brood box

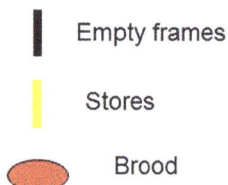

| Empty frames

| Stores

⬭ Brood

RESULT – brood in the shallow box emerges and it becomes the first honey super and the brood nest extends down into the new shallow box below

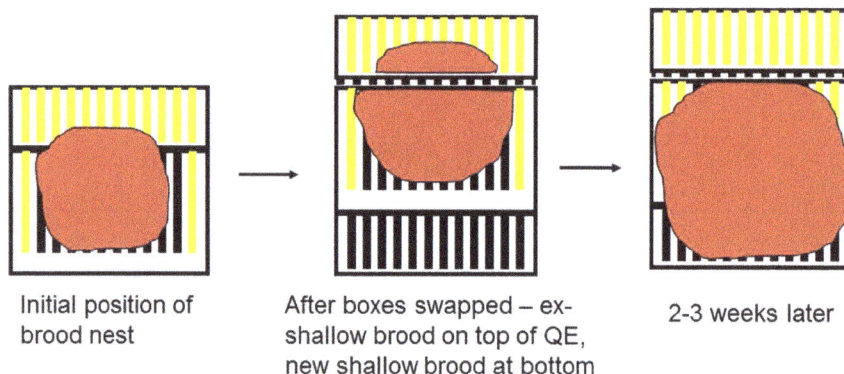

Initial position of brood nest

After boxes swapped – ex-shallow brood on top of QE, new shallow brood at bottom

2-3 weeks later

If at the end of the season, there is a shallow box at the bottom of the hive it is advisable to move it up on the top of the deep brood box before feeding. Due to contraction of the brood nest, at this time of the year it will usually contain little or no brood and probably no stores. On top of the deep brood it will be preferentially used to store feed in a position that will be readily accessible to the winter cluster.

Contrary to what is said in most beekeeping books, the best place to create space for the queen to lay is below the existing brood nest. A box of drawn comb placed on top of the brood nest will almost inevitably be treated like a super and substantially filled with honey. Although in a good nectar flow the 'receiver' bees often dump incoming nectar in empty cells below the brood nest this is only a temporary measure and it will quickly be transferred further up the hive. Capped honey below the brood nest will also be removed when time allows and this is the basis of the box re-location shown in example Figure 3b.

Caution – If the hive floor has a traditional depth of 21-22mm (solid or open-mesh) it will usually be necessary to remove brace-comb from the bottom bars of the box immediately above the floor - otherwise it will not fit in its new position and numerous bees will be squashed. Frames **must** be shaken free of bees before attempting this process. Using floors with the correct depth for open-mesh floors (9mm for bottom bee-space hives and 15mm for top bee-space hives) the frames should have little or no brace-comb and will not need any trimming.

Other Hive Configurations

In the case of a hive on double deep configuration, controlling brood nest position is simply a matter of moving as many frames of brood into the top box as possible (taking care to create a sensible nest shape that the nurse bees can cover efficiently) and moving frames of stores or empty comb down. Stores beneath the brood nest will quickly be removed and space will become available for the queen to lay. Because double brood provides more than enough space for most colonies the situation is less critical. However, the colony is likely to end the season with a substantial amount of honey beneath the queen excluder – which is good or bad depending on your point of view (ie. how much you want to feed).

Using a single deep brood box hive configuration (which gives less space than the potential of most queens), the position of the brood nest is less likely to present problems. Here it is just a matter of ensuring that, as far as possible, all the frames pull their weight (are available for the queen to lay on). To achieve this aim the frames should contain a minimum amount of honey and pollen stores. Well managed single deep box hives tend to have more pollen stored in the first super simply because there is nowhere else to put it and, by default, it becomes a brood-less extension of the brood nest.

Extra deep box hives (eg. 12x14) can have a problem with brood nest position and this is not quite so easy to remedy. **Figure 4** shows how this can be done by introducing frames to the middle of the box – but there must a good head of bees when this is attempted. Depending on what is available, these can either be frames of foundation, empty drawn frames or existing lateral frames (from the outside of the box) with any honey stores scored using an uncapping fork. The colony will want to restore brood nest integrity and these frames will quickly become part of a larger, spherical brood nest.

Figure 4 - **Expanding the brood nest in single box hives using a deep or extra deep box**

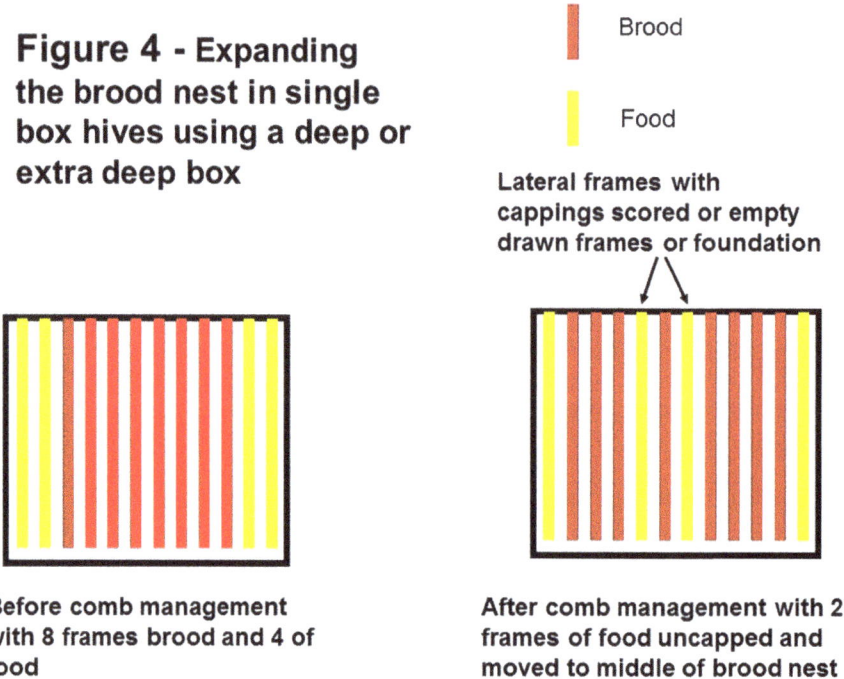

Brood

Food

Lateral frames with cappings scored or empty drawn frames or foundation

Before comb management with 8 frames brood and 4 of food

After comb management with 2 frames of food uncapped and moved to middle of brood nest

Supering

Also part of box management is adequate supering of a colony. It is essential to provide plenty of space for the processing of nectar and the storage of honey. It should be remembered that fresh nectar has 2-4 times the volume of the honey that it becomes and extra space should be provided for the bees to conduct the drying process. Supering needs to be kept one jump ahead of the need for storage but without creating so much volume that heat loss becomes a problem. When the top super is full of bees (wall-to-wall), regardless of the fact that it may be only part-full of honey, is the time to add the next super. However, no amount of supering will act as a substitute for poor brood nest management. Weather is the problem that the beekeeper can do nothing about and during prolonged adverse conditions the bees will inevitably move down from the supers and crowd the brood area and this will often trigger swarming.

1.3 Brood Relocation

This is one of the oldest tricks in the book and usually goes under the name of the Demaree Method, dating back to 1892 – a method that is probably under-used in modern beekeeping. There are many variations of the method but the basic principle is the removal of frames of brood from the bottom of the hive and relocating them in a new box at the top of the hive – above the supers. The removed frames are replaced by empty drawn frames (if not available foundation can be used) thus giving extra space for the queen to lay. Brood at the top of the hive attracts nurse bees to move up to cover it and this serves to reduce congestion at the bottom of the hive. The combination of new laying space for the queen and a reduction in congestion in the brood area inhibits the impulse to swarm.

The method was originally designed for hives on double (or triple) deep brood boxes and when brood has emerged from the frames put to the top the plan is to return them to the bottom of the hive in exchange for ones containing more recent brood (a frame circulation system).

Figure 5a shows a classical Demaree with two deep brood boxes. **Figure 5b** shows how the method can be applied to a hive on a single deep brood box. The manipulations shown in 5**b** can easily be adapted to brood and a half configuration – the shallow brood stays at the bottom of the hive, either above or below the deep brood box depending on the position of the brood nest (see **Section 1.2** above).

Running a hive on a single deep box, it is doubtful if the beekeeper would ever want to re-locate the entire brood nest to the top of the hive as this would seriously unbalance the colony. Typically only 4-8 frames are moved at any one time, resulting an incomplete Demaree box as shown in **Figure 5b**. An incomplete set of frames should be flanked by a dummy board on either side (colour coded blue). Further frames of brood can be moved up at a later date if required so it may eventually become a complete box. A part-filled Demaree box is no problem but the beekeeper needs to be aware that in good nectar flow wild comb may be built in any unoccupied space at the sides of the box.

Figure 5a - Classical Demaree applied to a hive on double brood with 14 deep frames of brood

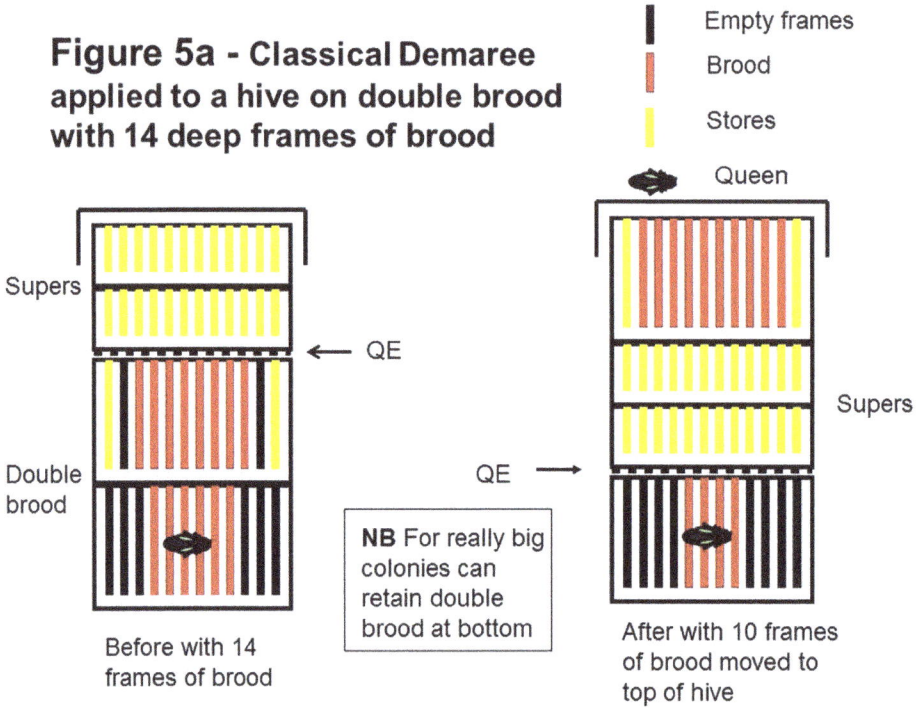

Empty frames

Brood

Stores

Queen

Supers

← QE

Double brood

QE →

NB For really big colonies can retain double brood at bottom

Supers

Before with 14 frames of brood

After with 10 frames of brood moved to top of hive

Figure 5b - Demaree applied to a hive on a single deep brood box with 12 frames of brood

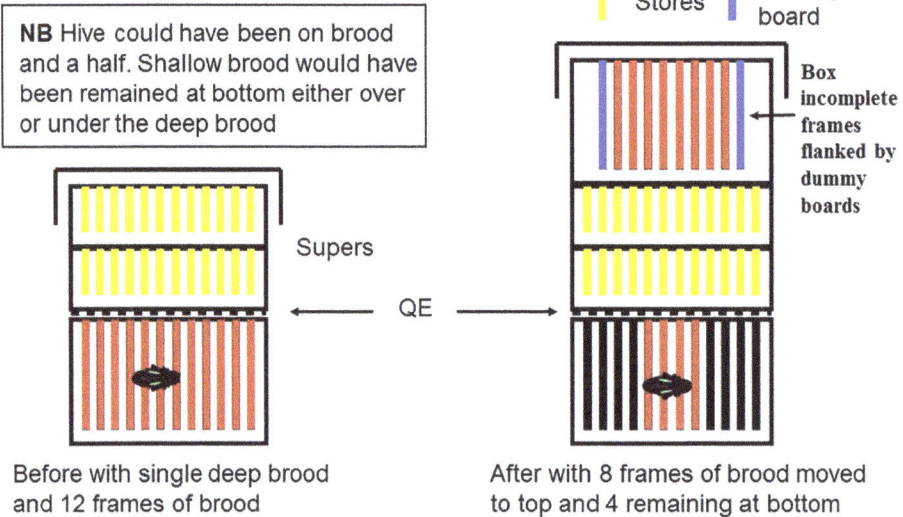

NB Hive could have been on brood and a half. Shallow brood would have been remained at bottom either over or under the deep brood

Empty frames

Brood

Stores

Dummy board

Box incomplete frames flanked by dummy boards

Supers

← QE →

Before with single deep brood and 12 frames of brood

After with 8 frames of brood moved to top and 4 remaining at bottom

The Demaree Method is quite an effective method of pre-emptive swarm control but it does have some downsides:-

▸ The first is that the bees covering the brood at the top of the hive may be far enough removed from the queen that they regard themselves as queen-less and start emergency queen cells. The greater the spatial separation (the number of supers) the more likely this is to happen. After 5-7 days it is necessary to carefully examine the top box for queen cells and destroy them.

▸ The second is that frame re-cycling is more difficult than appears at first sight because as soon as the brood has emerged the vacated cells are quickly filled with nectar and later with capped honey. This means that the top box normally has to be left in place for the rest of the season and removed as part of the honey harvest. However, providing the frames are fairly new this should not adversely affect honey quality.

1.4 Splitting Colonies

The previous methods of pre-emptive swarm control have kept the colony in one piece. Splitting is different and (potentially) creates a second colony. However, splitting is the most powerful and reliable method of pre-emptive control and has a long history of use in beekeeping. Splitting colonies is also a method of making increase which is covered in more detail in the WBKA booklet, *'Simple Methods of Making Increase'*.

It is often said that splitting a colony is the enemy of a good honey crop. However, splitting is always better than losing a swarm - unless you can guarantee catching it! The effect on the honey crop depends crucially on the timing of the split. If it is done at the right time, eg. directly after the spring flow, it can result in an enhanced yield. Under the right circumstances the two resultant colonies can produce more than the original (one) colony - even if it did not eventually swarm. A controlled split is always better than an artificial swarm because it enables the beekeeper to create a better age-class distribution of worker bees in both parts of the split.

When splitting a colony to provide pre-emptive swarm control the following considerations should be taken into account:-

▶ The split should be sufficiently radical to provide swarm control for the rest of the season.

▶ Both sides of the split should be viable, ie. adequate bees, brood and stores.

▶ The timing should be right for the colony, ie. its state of development.

▶ The timing should be right for potential nectar flows, ie. allowing time for the colonies to re-build.

No firm guidance (prescription) is possible and the details of splitting depend on the judgement of the beekeeper. **Figures 6a** (using a new hive stand) and **Figure 6b** (using a split board) are examples which illustrate the principle. In both Figures the blue box and the frames within are new.

Figure 6a – splitting a hive on brood and a half with 10 deep frames of brood, 7 of which placed new box on new hive stand

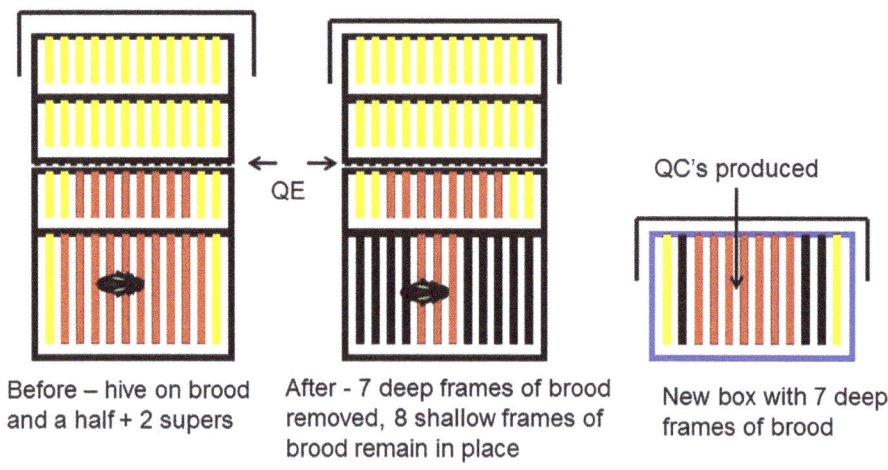

Empty comb
Brood
Stores
Queen

Before – hive on brood and a half + 2 supers

After - 7 deep frames of brood removed, 8 shallow frames of brood remain in place

QE

QC's produced

New box with 7 deep frames of brood

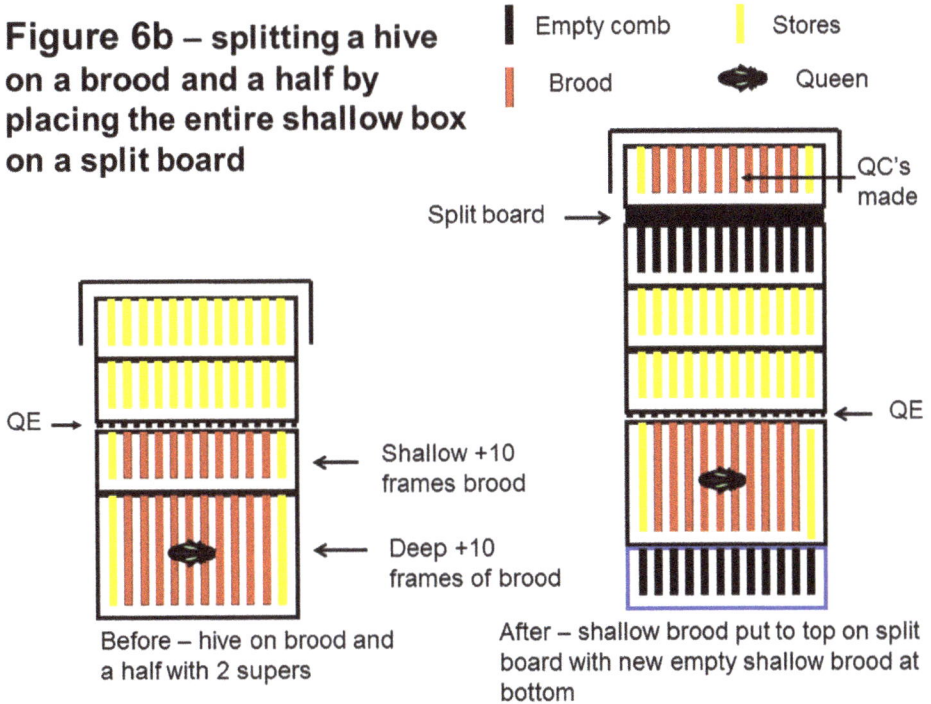

Figure 6b – splitting a hive on a brood and a half by placing the entire shallow box on a split board

Empty comb   Stores

Brood   Queen

Split board →

QC's made

QE →

← QE

QE →

← Shallow +10 frames brood

← Deep +10 frames of brood

Before – hive on brood and a half with 2 supers

After – shallow brood put to top on split board with new empty shallow brood at bottom

Recombining Splits

If a split has been made properly (see guidelines above) there is about a 90%+ probability that the queen-less part of the split will raise a new laying queen. The beekeeper now has two colonies instead of one and this may not be the desired outcome. Simple logic dictates that you cannot go on indefinitely doubling the number of your colonies so the solution is recombination (uniting colonies) but how and when? Early in the season recombination is not usually a sensible option as it will only create a large colony that is likely to undo all your good work and swarm. Later in the season, when the swarming impulse is on the decline, it is possible to recombine to produce a 'super-colony' for the main honey flow or for taking to the heather.

There are three main options for the new colony (with a new queen) which will currently be living either on a new stand or a split board.

▶ Giving it permanent independence, ie creating a new long-term colony.

▶ Recombining later in the season to produce a 'super-colony' (with a choice of queens to be made by the beekeeper).

▶ Giving it temporary independence, ie. supering it, keeping until the end of the season and then recombining (also with a choice of queens).

The word 'recombination' implies that the new colony has to be united with the original colony. However, much greater flexibility is possible and it is often the preferred option to unite it with a colony that has an old or unsatisfactory queen as a simple method of re-queening. Artificial swarming also creates two colonies from one and there is further discussion about how to deal with this situation in **Part 2 (Aftermath of Artificial Swarming)**.

PART 2 – RE-ACTIVE SWARM CONTROL

Re-active swarm control starts where pre-emptive swarm control fails - when queen cells with contents are found in the hive. When this happens there is no known (reliable) method by which the swarming process can be turned off and the beekeeper needs to accept the fact that unless something is done to prevent it, that colony will inevitably swarm. Destroying queen cells only delays swarming and may make the situation worse, ie. the colony swarms with barely-started second generation queen cells and takes a higher proportion of the bees with it. Worse still is destroying queen cells not realising that the colony has already swarmed. If swarming occurred more then 4-5 days earlier there will be no eggs and young larvae present from which emergency queen cells can be made and the colony will be rendered queen-less.

Determining the Stage in the Swarming Process

On finding queen cells in a colony the first thing to do is to determine what stage in the swarming process the colony has reached. However, before jumping to conclusions, the beekeeper needs to be sure that the queen cells are actually **swarm cell**s and not supersedure or emergency queen cells. This is usually obvious by their position and number but, if in doubt, consult WBKA booklet, *'There are queen cells in my hive – what should I do?'*

There are four main stages in the swarming process, each with a different solution:-

1. **Colony has queen cells but has not yet swarmed.**

2. **Colony has issued the prime swarm but has not yet cast swarm.**

3. **Colony has emerged queen cells and may (or may not) have issued a cast swarm.**

4. **There is evidence that the colony has swarmed but it currently appears to be queen-less (ie. it has no brood of any age), and beekeeper has no idea what happened and when.**

How to Investigate

Full details of how to do this are given in the WBKA booklet which contains a diagnostic tree with 12 steps, each consisting of the method of **Investigation** and the recommended **Remedial Action**.

The beekeeper should do **ABSOLUTELY NOTHING** until a careful investigation has been completed. Fortunately the key information persists in the hive (on the frames) until at least 3 weeks after the prime swarm has departed – the last time the queen can have laid some eggs.

2.1 Colony has not yet Swarmed

How do you know? – the evidence in order of reliability:-

- Queen seen

- Recently laid eggs (standing on end).

- Eggs leaning over at about 45° were laid 24-48 hour previously and those lying flat 48 hours plus and about to hatch.

- Be aware that a late afternoon inspection may reveal standing –up eggs but a swarm could have departed a few hours earlier

- No missing bees (colony the size you expect)*

- Maturity of queen cells and recent weather – has it been conducive to swarming?

*NB – No missing bees is to be expected if the queen has had her wings clipped - the colony can have swarmed and there will be little or no loss of bees. This can also occur naturally if the queen is for some reason unable to fly – **so beware of this not uncommon possibility**.

The beekeeper really needs to see the queen or recently laid eggs early in the day to be absolutely sure. If the colony is deemed not to have swarmed the remedy is to make an **artificial swarm**.

Artificial Swarming

Having established that the colony has not yet swarmed the (only) solution is artificial swarming. Most beekeeping books describe what is called the Pagden Method, first published in 1868. The basic principle of this method is that the **parent colony**, comprising the brood and queen cells, is separated from the **artificial swarm**, comprising the queen and the flying bees. The artificial swarm remains in the same location and the parent colony is moved to a new one (nearby). Our experience over about 15 years of trying to make this method work showed that it had an unacceptably high failure rate – probably in excess of 50%. By 'failure' what is meant is that the artificial swarm does not reliably lose the impulse to swarm and can resume its original plan (to swarm) at more or less any time during the next 3 weeks. Further discussion of the Pagden Method can be found in **Appendix 1**

The method described in this booklet has a virtually 100% success rate and is called Snelgrove II (modified). The fundamental difference between this and the Pagden method is that the queen does not go to the artificial swarm but remains in the parent colony along with the brood and queen cells. Many beekeepers, brought-up on the Pagden Method, find this counter-intuitive and do not trust having the queen with brood and queen cells. The simple solution is to try it and then you will know.

As the name suggests, the basic method originates with L.E. Snelgrove, so his name has been retained out of respect for a very distinguished beekeeper. The modified version is subtlty different by providing the artificial swarm with the means of making emergency queen cells. This is thought to be the secret of its success by switching off (or effectively suppressing) the swarming impulse. Despite the name it does not require the use of a Snelgrove board (or a split board). It can also be done with the parent colony moved to a separate stand but use of a split board does have many advantages (see discussion below).

Artificial swarming by the Snelgrove II (mod) Method is a two-stage operation which is illustrated in **Figures 7a and b.** using a new hive stand (not a split board in sight!). For clarity the brood area of the **parent colony** is in blue boxes (with a blue background) and the **artificial swarm** in a green box (with a green background).

Figure 7a - Snelgrove II (modified) using separate hive stand - the initial manipulation

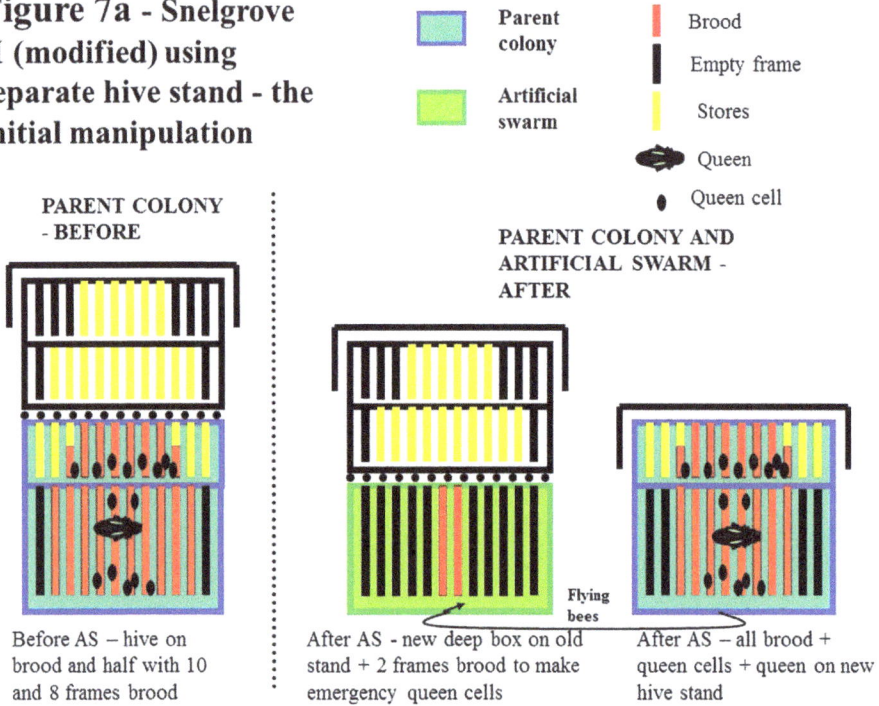

Legend:
- Parent colony
- Artificial swarm
- Brood
- Empty frame
- Stores
- Queen
- Queen cell

PARENT COLONY - BEFORE

PARENT COLONY AND ARTIFICIAL SWARM - AFTER

Flying bees

Before AS – hive on brood and half with 10 and 8 frames brood

After AS - new deep box on old stand + 2 frames brood to make emergency queen cells

After AS – all brood + queen cells + queen on new hive stand

In the **initial manipulation** (**Figure 7a**) all the brood, including queen and queen cells, are moved on their existing floor to a new hive stand within the apiary - reasonably close is convenient but more than 3 feet apart is essential. This is now referred to as the **Parent Colony**. A new floor is placed on the vacated hive stand and a new box with 10 frames of preferably drawn comb (but a mixture of drawn and foundation will suffice) is placed on it. This will become the **Artificial Swarm**. Two bee-free frames of brood are now removed from the parent colony and transferred to the middle of the artificial swarm. These two frames **must** contain eggs and young larvae from which emergency queen cells can be made but **must not** have any queen cells (or the queen) on them. The missing frames in the parent colony should be replaced using drawn frames if possible.

Both hives are now re-built and most (or all) of the supers are normally given to the artificial swarm where most of the current foraging force will reside. However, remembering that the parent colony is initially going to lose all its flying bees, the quantity of available stores needs to be taken into consideration. If these are thought to be insufficient to last until a new foraging force has developed or there is unlikely to be a significant nectar flow in the near future, the parent colony should be given a queen excluder and one (or possibly more) of the supers. In some seasons it may even have to be fed.

The first thing that happens is that the parent colony on the adjacent stand will lose its flying bees back to the artificial swarm. Amongst these will be the bees that are running the swarming process. For this reason, the parent colony entirely loses the impulse to swarm and in due course the queen cells will be torn down and the queen will resume laying. In the queen-less artificial swarm the bees will start emergency queen cells – and this appears to be vital to the loss of the swarming impulse.

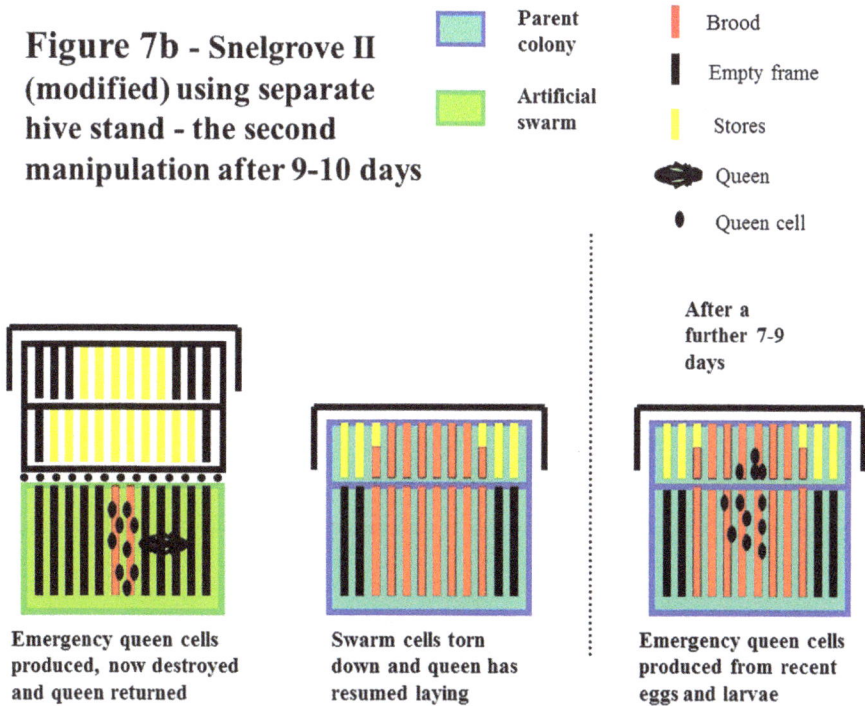

Figure 7b - Snelgrove II (modified) using separate hive stand - the second manipulation after 9-10 days

Parent colony

Artificial swarm

Brood

Empty frame

Stores

Queen

Queen cell

After a further 7-9 days

Emergency queen cells produced, now destroyed and queen returned

Swarm cells torn down and queen has resumed laying

Emergency queen cells produced from recent eggs and larvae

The **second manipulation** should take place 9-10 days later (12 days is the absolute safe limit). Timing is important because all the queen cells in the parent colony should be torn down by this time. Usually this happens quite quickly but it does depend on the maturity of the queen cells, ie. it is not done until the virgin queens they contain are 3-4 days from emergence at which time the colony becomes fully aware of their presence. However, if the queen cells were only just started at the time of the split it may take a day or two longer. More importantly, the second manipulation must be done before any virgin queens can emerge into the artificial swarm. The calculation goes like this - if an emergency queen cell is based on a 1 day old larva (day 4 from when the egg was laid) the first queen could emerge sometime on day 12. From this it can be seen that day 12 is pushing the limit and it is better not to take the risk.

The second manipulation starts with checking that the queen cells in the parent colony have been torn down and the queen has resumed laying. If this has not happened it means that there is **no** queen in the parent colony. The most likely cause for this is that the beekeeper failed to notice that the colony had already swarmed (sometimes it is difficult to be sure). This is not a cause for alarm because the parent colony has queen cells from which it can re-queen itself (or may by this time contain a virgin queen) and the artificial swarm has emergency queen cells that are approaching maturity. The parent colony will re-queen without swarming (no management required) but, the artificial swarm may need steps to be taken to prevent cast swarming (see **Section 2.2 below**). There are other less likely explanations for the queen cells in the parent colony not having been torn down and the beekeeper should refer to the **Fault Finder** in **Appendix 2** to understand the situation and what to do next. The next step is to examine the artificial swarm where the two frames with eggs and young brood on them should now have emergency queen cells on them. If this is not the case the queen (or a queen) must be present somewhere in the artificial swarm. Again the beekeeper should refer to **Appendix 2**.

The frames with queen cells should now be removed and then the queen can be transferred (repatriated) to the artificial swarm. Details of the best way to do this (smoothly and safely) can also be found in **Appendix 3.**

The question now is what to do with the two frames with emergency queen cells on them? If at the time the colony was artificially swarmed there were plenty of younger bees in the supers (and this is usually the case with a colony that has set up to swarm) then the occupants of the emergency queen cells will have been well-fed and contain fully developed virgin queens. If the beekeeper is confident this is the case, the two frames with emergency queen cells can be placed in the **parent colony** to provide it with a new queen. If there is any doubt about the quality of the cells then it is better to destroy them and let the **parent colony** start from scratch and make its own. It will have available suitable brood produced by the queen who had resumed laying. Also there will be plenty of nurse bees from the still emerging brood to feed them well. The latter is the safer option but the former saves 9-10 days on re-queening of the parent colony. Which option to take requires a judgement call to be made by the beekeeper.

Starting with an identical hive that has set up to swarm, **Figures 8a and b** show the equivalent process using a spilt board (colour coding of boxes the same as Figures 7a and b). The overall process is just the same but repatriation of the queen in the second manipulation requires slightly different logistics which are described in **Appendix 3.**

WARNING – If when using a split board, the artificial swarm and the parent colony are (vertically) too close together there is a small chance that the (queen-less) bees in the artificial swarm will find that their 'beloved 'queen has just moved 'up-stairs'. If they do find her – and they can initially be seen following a scent trail up the side of the hive to the entrance on the split board – the parent colony will inevitably issue a swarm in the next few days. If the bees in the artificial swarm have found the queen the only remedy is to immediately move the parent colony to a new position in the apiary so that contact is broken.

The first reaction of a colony on losing its queen is to frantically look for her. If she cannot be found they will then start the emergency queen cells. Obviously if the vertical separation is small, eg. one super's worth, they are more likely to find her. Two supers is safer and three is secure. Good pre-emptive swarm control usually means that the colony will be well-developed at the time it sets up to swarm and will have three (or more) supers on it but, if this is not the case, then extra supers should be added to enable greater separation to be given. The alternative is to use the two stand version (**Figures 7a and b**), where the artificial swarm (more than 3 feet away) will be unable to find where the queen has gone.

Figure 8a - Snelgrove II (modified) using a split board - the initial manipulation

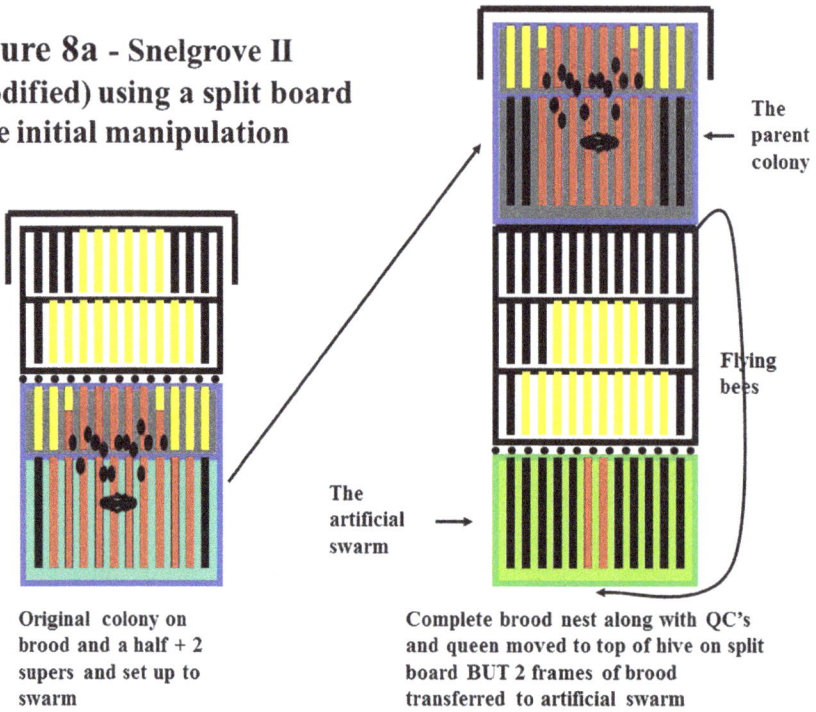

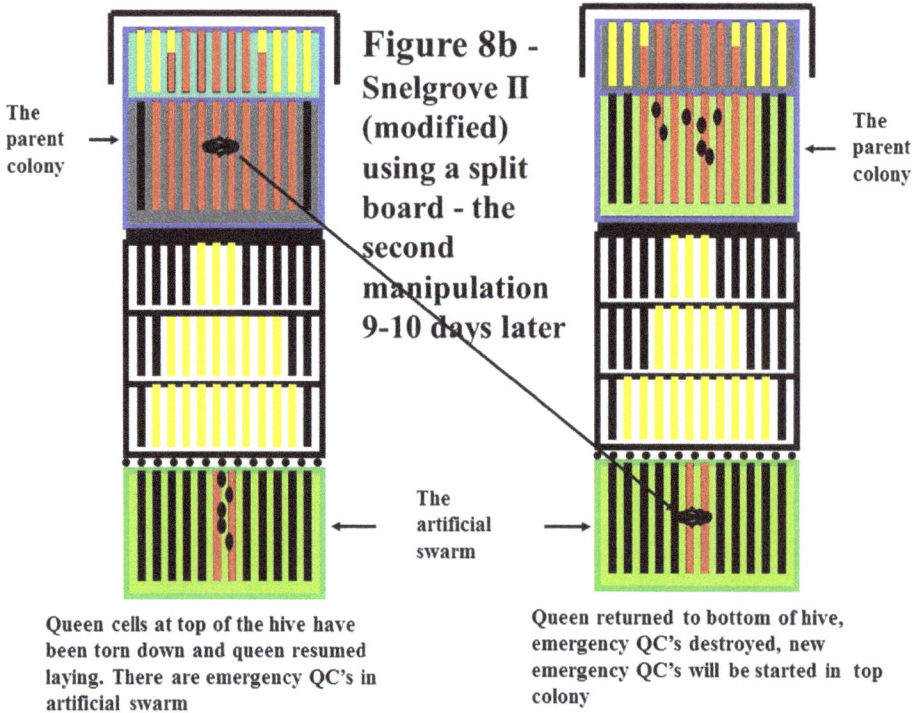

The parent colony

Flying bees

The artificial swarm

Original colony on brood and a half + 2 supers and set up to swarm

Complete brood nest along with QC's and queen moved to top of hive on split board BUT 2 frames of brood transferred to artificial swarm

Figure 8b - Snelgrove II (modified) using a split board - the second manipulation 9-10 days later

The parent colony

The parent colony

The artificial swarm

The artificial swarm

Queen cells at top of the hive have been torn down and queen resumed laying. There are emergency QC's in artificial swarm

Queen returned to bottom of hive, emergency QC's destroyed, new emergency QC's will be started in top colony

Advantages of Snelgrove II (modified)

Apart from its virtually 100% success rate, one of the main advantages of this method of artificial swarming is that for the first manipulation the queen does not have to be found. Even for an experienced beekeeper, this is often not easy when the colony is crammed with bees and the queen is being constantly harassed by the workers to slim her down. She could be anywhere and probably running as fast as she can. When in the second manipulation she **does** have to be found there will be fewer bees and she will have settled back to the day job (laying eggs) – so finding her should be much easier.

Got the timing wrong?

If, for some reason, the beekeeper gets the time of the second manipulation wrong, and the emergency queen cells have already released virgins into the artificial swarm, it is not the end of the world. However, it is unlikely that all the virgin queens can be found and the old queen safely repatriated and it is better to allow the artificial swarm to re-queen naturally and leave the old queen where she is in the parent colony. The latter can now be given a queen excluder and supers and run as the main-honey producing unit.

However, there is conflicting testimony from users as to what happens to the artificial swarm which has multiple queen cells. Does it behave in a normal way to a colony that has made emergency queen cells and select a new queen without swarming? Or finding itself with multiple queens, does it proceed to issue a swarm (or swarms)? Or are both outcomes possible? At the present time this is not known and a safe way of dealing with this uncertainty is given in **Section 2.3** (below).

Aftermath of Artificial Swarming

If successful (ie. the new queen has mated and started laying), when the process of artificial swarming is complete there will be two colonies instead of one and this may or may not be what the beekeeper wants.

▶ If an increase in the number of colonies is not required then re-uniting is the answer. This is easier when using a split board than it is with the new colony on a separate hive stand. Re-uniting should not (normally) be attempted until the parent colony has developed a new laying queen and she has proven herself to be a good'un. Using a split board, a two-tier colony (one on top of the other each with its own supers) can be retained until the end of the season when a decision as to its future can be made. **Note** - this is not a two-queen system because the two colonies are independent and have different entrances and separate supers. When uniting two colonies it is preferable for the beekeeper to choose the queen because letting the colonies do it for themselves **can** result in serious conflict. Also I find it difficult to believe the old adage that 'the bees will always choose the best queen'. From the beekeeper's point of view the new queen is usually preferable.

▶ If extra colonies are required then, with the new colony on a separate stand, this has already been accomplished. If a split board has been used then moving the new colony to a different apiary to give it independence is preferable as there will be no loss of flying bees. If a new colony on a split board is to be kept in the same apiary care must be taken to avoid excessive loss of flying bees. It is recommended that independence-day should be delayed for a minimum of 5-6 weeks after the new queen has started to lay – which, in reality, may be pretty much the end of the season.

2.2 Colony has already issued the prime swarm but has not cast swarm

It is assumed that the beekeeper has already determined that the prime swarm has departed with the old queen. It is then just a matter of carefully checking that no queen cells have emerged. The age of the youngest brood gives some clues, eg. if there are no eggs but some newly hatched larvae then the swarm occurred 3-4 days ago and, under normal circumstances, the queen cells are about 4 days from emergence.

In this situation, unless something is done to prevent it, the colony is likely to issue at least one cast swarm. The timing of this swarm will be 2-4 days after the first virgin queen cell has emerged. There are two methods of preventing this:-

1)   Thinning the queen cells to just one, selecting a cell of good size in a well-protected position. If there are still eggs or young larvae present it is best to delay this operation until they are past the stage when these could be used to make emergency queen cells. Alternatively the existing queen cells can be thinned now and the hive checked for new queen cells in a few days' time.

2)   The queen cells can be left intact until the estimated date for their emergence – a few queen cells can be investigated to check on their state of development and likely due-date. What to do next is described in the next section (2.3).

2.3 Colony has emerged queen cells and may (or may not) have issued a cast swarm

There is no easy way of knowing whether the colony has already cast swarmed and no way of knowing for sure if the colony contains an emerged virgin queen (or queens). Most of the remaining queen cells will contain queens that are waiting to emerge but are being prevented from doing so by 'warder bees' posted on the cells. Using the tip of a knife the beekeeper should carefully open several cells and let the virgin queens walk out into the colony. When you think you have done enough of this (ie. given the bees plenty to choose from), **ALL** remaining queen cells must be destroyed. It does not seem to matter how many virgin queens you release in this way, the colony will proceed to select one (by whatever means) and make no attempt to swarm. This is mission accomplished - the bees have selected the new queen **not you**. Further information can be found in the WBKA booklet, *'There are queen cells in my hive – what should I do?'*

2.4 Colony appears to be queen-less (it has no brood) and beekeeper has no idea what happened and when

There may be a queen in the colony who is just about to start laying but, without actually finding her, you cannot tell. The behaviour of the colony (the bees seem calm) and the presence of laying arcs (cells prepared for a queen to lay) imply that all is well with the colony but neither of these signs is completely reliable. The best thing to do in this situation is to insert a test frame (taken from another hive) containing eggs and young larvae. If the colony is queen-less emergency queen cells will be made on this frame but if they think they have a queen the donor brood will be raised in the normal way. Even this is not 100% reliable because in rare cases the colony may contain a non-laying queen and no further progress can be made (ie. the colony cannot be re-queened) until she has been found and removed. Further information on this situation can be found in the WBKA booklet, *'There are queen cells in my hive – what should I do?'*

2.5 Late Season Swarming

This can be one of the unfortunate by-products of pre-emptive swarm control. Typically a large colony that has been kept together with no attempt at swarming until late-June or early-July will suddenly develop queen cells. This could still be controlled by means of artificial swarming (as described above) but splitting the colony just when the main nectar flow is about to start is definitely the last thing the beekeeper wants. One method of dealing with this situation is by simply removing the queen. You do not need to do anything drastic like kill her (you will probably want to repatriate her later), you merely put her aside to tick-over with a few workers in support. We call this practice putting the queen in 'purdah'; it is not a perfect solution but it works well enough.

Figure 9 - Late season method of dealing with colony set up to swarm putting by the queen in 'Purdah'

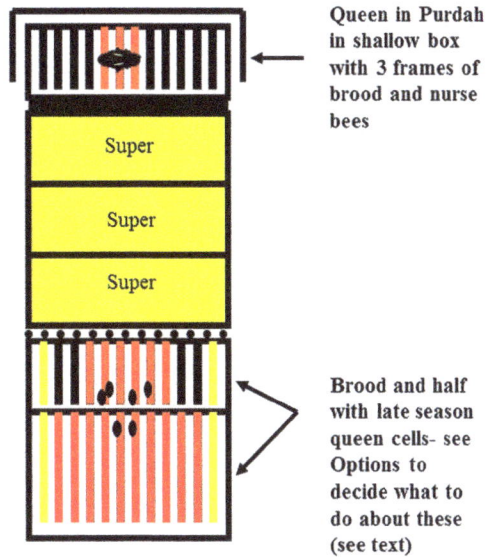

Queen in Purdah in shallow box with 3 frames of brood and nurse bees

Super

Super

Super

Brood and half with late season queen cells- see Options to decide what to do about these (see text)

As soon as queen cells are found the queen should be removed along with a small amount of brood and some worker bees and either installed in a nuc box or a shallow brood box which is placed on a split board on top of the hive. There must be sufficient bees to support her and cover any brood (remember all the flying bees will return to the bottom of the hive). There also needs to be enough stores because foraging activity will be limited. It is also a time of year when robbing may occur so there needs to be enough bees and a small entrance to prevent this happening.

What to do with the queen cells?

The best option is to destroy the queen cells but first ensuring that there are eggs and young larvae present from which the colony can start emergency queen cells. The colony will probably be switched into emergency re-queening mode and when these cells are mature (about 12 days later) they will choose a new queen and not swarm. If you want to be sure that swarming does not occur then follow the method of releasing virgin queens as described in **Section 2.3** (above).

BE AWARE - queens mated later in the season (from July on) are known to be less reliable (less likely to survive their first winter) than those mated earlier (May and June). It is thought this is due to the increased probability of the queen mating with a drone(s) that is carrying deformed wing virus (DWV) and becoming infected herself – in other words a sexually transmitted disease (STD). For this reason 9-10 days later it is better to destroy the emergency queen cells and repatriate the old queen, as in the second manipulation for the Snelgrove II method described above.

During this enforced queen-less period the colony will continue to forage well (but probably not quite as well as it would with an incumbent laying queen) and there will be little loss of honey crop.

A new method of dealing with a late season swarming colony

Since the first edition of this booklet was written we have been experimenting with methods of dealing with late swarming and so far (the last 3 years with more than 20 replications) the method shown in **Figure 10** (the Demaree Method) has been completely reliable. As shown in Figure 10, the colony is in the same condition as it would be for the second manipulation of Snelgrove II (mod) – the queen cells in the parent colony have been torn down and the queen has resumed laying. There are emergency queen cells in the artificial swarm. Initially the manipulation proceeds the same way but this time destroying the emergency queen cells before the queen is repatriated to the artificial swarm.

The next move makes use of the multiple doors on the Snelgrove board but a single entrance split board can simply be turned so the entrance is in the required new position. First the mesh from the middle of the board is removed so the parent colony and artificial swarm can mingle. The door on the Snelgrove from which the bees have been flying is closed and door at the front of the hive (top or bottom or both) is opened (or the single entrance board is turned to this position). The bees that fly from the parent colony will soon adapt to the new orientation (a 90° change makes for an easier transition). After 24 hours the board should be removed altogether and the bees will quickly adapt to use the entrance at the bottom of the hive.

NB. It is essential that the bees in the parent colony are forced to descend to the box where the queen now is in order to fly. If this does not happen they will think they are queen–less and start emergency queen cells at the top of the hive.

Figure 10 - Demaree Method of re-uniting a late swarming colony

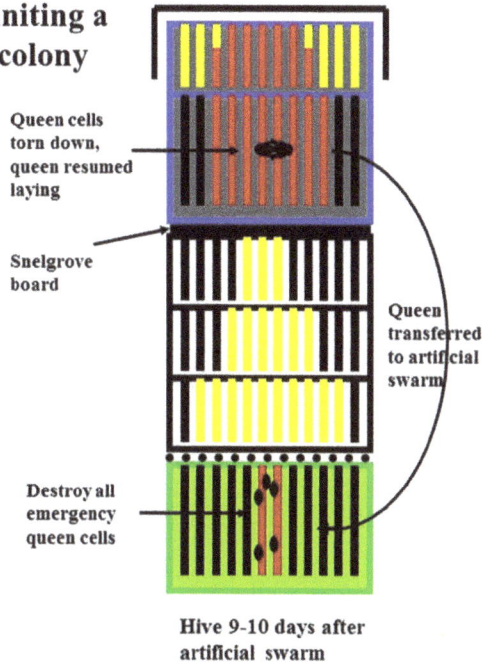

Queen cells torn down, queen resumed laying

Snelgrove board

Queen transferred to artificial swarm

Destroy all emergency queen cells

Hive 9-10 days after artificial swarm

The desired result has now been achieved and, after limited separation (9-10 days), the whole colony (and particularly the foraging force) is back together as one unit. Brood in the Demaree boxes will emerge and, depending on the time and the nectar flow, the frames will be used to store honey which can become part of the honey harvest (or later used as food frames for the bottom brood area).

Postscript

The above methods of pre-emptive and re-active swarm control should provide the beekeeper with a comprehensive package of management techniques by which the swarming impulse of honey bee colonies can be kept under some measure of control. The qualification 'some measure' is used because the beekeeper who claims to have achieved complete control is probably 'not of this world' or is being economical with the truth. Colonies are individuals and the most common type to evade swarm control are those that start queen cells and swarm long before the first one is sealed, thereby evading detection by routine inspection (a **Houdini** colony). Even with a 6-7 day inspection interval during the swarming season it is impossible to prevent this sort of thing from occasionally happening – except by clipping the queen's wings. Queen clipping is not really a method of swarm control, it just extends the safe inspection period to about 14 days. As a management practice it has pros and cons and often elicits strong opinions. A full discussion of queen clipping is beyond the scope of this booklet.

To bring this booklet through the full circle, I return to what was said at the end of the second paragraph of the Introduction; '*there is no doubt that swarm control is simultaneously the most important and most difficult aspect of colony management'.* We can but try!

APPENDIX 1. –
Further Discussion of the
Pagden Method

Because this is the most widely recommended method of artificial swarming which can be found in most beekeeping books (usually with little explanation), it may be worth some further discussion.

The theory behind the Pagden Method is that the artificial swarm it creates resembles a natural swarm, 'thinks' it **has** swarmed and will now settle down to life in the slow lane. The reality is that the artificial swarm is nothing like a natural swarm because it has an entirely different age structure, being dominated by older bees, whereas a natural swarm is dominated by young bees. The subsequent performance of an artificial swarm compared with that of a natural swarm clearly underlines the difference.

The probable explanation for the unreliability of the Pagden Method is because the artificial swarm created by the beekeeper has a structure that does not occur in nature. This means that the colony has no evolved (hardwired) solution in its behaviour repertoire that it can engage in order to achieve the best outcome (for itself). The only possible options are either to start some new queen cells and persist with the swarming impulse or abandon the idea, settle down and re-build a more normal age structure as quickly as possible to ensure survival. Users of the method obviously expect/hope that the artificial swarm will adopt the second option. The Heddon modification does not influence the outcome and probably only makes the situation worst by transferring more bees to the artificial swarm, thus making the potential swarm larger. The decision (to swarm or not to swarm) seems to be finely balanced and there is no known way the beekeeper can influence this.

The method described in the main text, Snelrove II (modified), differs by putting the artificial swarm into a condition which it **does understand** and **does** have a clear-cut response to ensure its survival, ie. it finds itself queen-less but with the means to make a replacement (emergency) queen. Being forced to make emergency queen cells puts the colony into emergency re-queening mode and this switches off (or suppresses) the swarming impulse.

APPENDIX 2. Fault Finding

Feedback since 1st edition - A few people who have used the Snelgrove II (mod.) Method have reported that it has not worked for them. With one exception (details below) they have been unable to provide sufficient details to determine the reason for the failure. Over the years we have had one failure in well over 100 replications but are confident we understand the cause (details below). However, there are other *possibilities and the following is a Fault Finding check list. **NB.** These are just logical possibilities and we do not know for sure whether anybody has ever made any of these mistakes.

1) As noted in the main text, the second manipulation (9-10 days later) starts with checking that the queen cells in the **parent colony** have been torn down and the queen has resumed laying. If the queen cells have **not** been torn down and there are **no eggs** to be found then a mistake was made in the initial manipulation, ie. the queen is **not** in the parent colony and she has either been lost altogether or is still somewhere in the artificial swarm.

2) The next step is to check the artificial swarm it see if emergency queen cells have been produced on the two donor frames. If there are queen cells then the missing queen cannot be in the artificial swarm. As already noted, the most likely cause is that the colony had already swarmed but there are other possibilities - the queen had been killed, dropped on the ground or has flown off.

3) If there are no queen cells then the queen is still residing somewhere in the artificial swarm. Another possibility is that the frames did not contain brood young enough for emergency queen cell to made.

4) If there are eggs and young larvae in brood box containing the artificial swarm then you have inadvertently created a Pagden type artificial swarm. In which case it will probably either have worked or not worked by now (9-10 days later).

   a) If it has swarmed then there should be no queen present and the age of the youngest brood will tell you when this happened.

b) If it has not swarmed (but still could do so) there will be a queen and recently laid eggs but you will need to check for the start of queen cells for about another 10 days.

c) In both cases (a and b) to prevent further swarming you need to thin queen cells or emerge virgin queens as recommended in **Section 2.3** in the main text.

5) If there are no queen cells in the artificial swarm there is one final possibility and this is that the queen somehow found her way into supers - but this must have happened recently or there would have been no queen cells in the brood box. If the queen is in the supers the best option is probably to transfer her to the parent colony and hope for the best.

Known Causes of Failure

Our failure – at the time the artificial swarm was made the colony was almost certainly on the verge of swarming. Either there was insufficient time for the flying bees to return to the artificial swarm or the weather prevented this happening. With the flying bees still present, the parent colony retained the urge to swarm and did so at the first opportunity.

Other known failure – being nervous about having queen cells and the queen together in the parent colony, one beekeeper destroyed all the queen cells himself instead of allowing the bees to do this in their own time. This altered the behaviour of the parent colony which swarmed at some time during the next few days.

APPENDIX 3. –
Recommended Logistics for the Second manipulation

If the queen cells in the parent colony have been torn down and the queen has resumed laying there will automatically be emergency queen cells in the artificial swarm. This is a sign that the first manipulation has gone according to plan. The main aim of the second manipulation is to find the queen in the parent colony and transfer to the artificial swarm.

If a split board has been used, the artificial swarm cannot be accessed until the parent colony (on its split board) has been removed from the top of the hive. It is best if the queen is found and put in a safe place before this is done. If the hive is high (and it usually is by now) it can be difficult to remove the top colony single-handed and it is useful to have an assistant. Having a spare roof or box available is a useful place to temporarily put the parent colony still on its split board.

It is always best to find the queen before there has been too much disturbance. In order to do this operation smoothly, the frame she is on should be removed and carefully placed in a holding box where she will be safe. A nuc. box (with its entrance blocked) can be used but we have a purpose-built box that holds a single frame specifically made for any manipulation that requires the queen to be put in a safe place (**see Figure 11**). This box comes in useful for other manipulations where you want to ensure the safety of the queen.

It is best to repatriate the queen on the frame where she is found surrounded by her own nest mates. She can be introduced on her own, and we have done this with no problem a few times. Other people have expressed concern about this practice (even reported that the queen had been killed) so, to avoid that possibility, this is a safer way to ensure her acceptance by bees from which she has been separated for 9-10 days.* Using a split board is inherently safer because the common hive smell is retained.

Figure 11 - Box for holding queen in safe place

Castelation to hold frame securely

Stabilising feet

Roof

SIN BIN

*By this time the emergency queen cells (that have just been removed from the artificial swarm) will have reached the stage in their development where the bees are fully aware of their presence and have started to show a keen interest in them. The colony should be expecting the emergence of a virgin queen in the near future but seems to quite happily accept that instead they suddenly get the gift of a mated and laying queen (a miracle?).

If the artificial swarm and the parent colony are on different stands Stage 2 is a relatively easy operation because both hives can be open at the same time. Having found and secured the queen, the next step is to remove the two frames with queen cells from the artificial swarm (supers removed first obviously). The frame with the queen on should then be placed in the gap thus created. Another frame from the parent colony, complete with its bees, can then be inserted next to it to fill the second space thus providing more 'friendly faces' nearby - and also a useful injection of nurse age bees.

It is then just a matter of deciding what to with the emergency queen cells; whether to destroy them and let the parent colony re-queen itself from scratch or transfer them intact to the parent colony to provide a new queen 9-10 days earlier (see discussion page 26).

The last task is to re-assemble the colonies complete with supers and adding extra supers if required. Bear in mind that the colony on the split board has yet to re-queen and the hive should be disturbed as little as possible until there is evidence that the new queen has started to lay. However, if access to the bottom part of the hive is for some reason required it should be done outside potential mating hours, ie. before 10am or after 5pm.

www.ingramcontent.com/pod-product-compliance
Lightning Source LLC
Chambersburg PA
CBHW061457270326
41931CB00021BA/3489